RECRUITING

Ryan Breslow

ISBN: 9798781807697

CONTENTS

This book is dedicated to everyone
trying to do something positive in the world.

1

INTRODUCTION

Many of us forget to put ourselves in others' shoes when we want something from someone. It's a simple thing—but easy to forget. We've all had the experience of looking for a job. So, before we get into the details of how to attract candidates to your company, let's turn it around. What do *you* look for in a role?

There are the basic answers: compensation, work-life balance, location, and so on. These are, of course, important. But let's assume all those factors are equal across a number of companies that have each made you an offer. In that case, what might separate one from the next? All the magic of recruiting lies in the answer to this question.

You'll notice already that none of this is rocket science. If you're looking for a book that's going to impart magic recruiting tricks, you've come to the wrong place. Your superpowers are unlocked by taking advantage of simple things, having discipline, and leaving no stone unturned. I'll be your guide.

Or, more precisely: I'll try to be the guide I *wish* I had had when I was starting Bolt. For the first six months, I was coding full time—spending all day behind a keyboard. I took pride in telling my friends how much I was coding. It sounds badass, right? It wasn't. As an army of one, it was impossible for me to focus on anything outside the codebase. User acquisition, building culture, fundraising

—none were possible because I needed to ship features.

One day, dissatisfied with my progress, I tried a different approach. It started with raising some money, which was exciting. But that didn't come close to the excitement I felt when we made our first engineering hire, the best day in Bolt history. This engineer was better than I was. They were shipping features faster. And, best of all, they were doing it without me. My mindset shifted: No longer would I take pride in being a great engineer—I would take pride in being a great recruiter of engineers.

Soon, we hired more world-class engineers to the team. VCs were impressed with the level of talent we were able to attract—preeminent engineers out of Facebook, Twitter, Google, Dropbox, and the like—and they gave us more money. This enabled us to hire even more engineers. Our product velocity was reaching new heights with every passing month.

We were off to the races—and I had learned that recruiting is simultaneously the most challenging and most rewarding part of company-building. It is the engine that drives all else. People are the heartbeat of any organization—and because they are, you must hire like your life depends on it.

Granted, Bolt faced challenges that not every company will face. We had to overcome immense technical hurdles and build hundreds of integrations across shopping carts, payment processors, tax and shipping systems, ERPs, and more to enable working checkout flows for online retailers. Not all businesses require this level of technical hiring upfront.

Still, hiring played a big role beyond engineering. Our first sales hire fell flat, unable to follow the sales playbook we had carefully developed. We were devastated, second-guessing whether our approach could actually scale. Then, we found a second sales hire: someone more senior—but still scrappy. His name was Mark Burner, and Burner was as smooth as butter! Still, Burner bombed in his first demo.

Devastated again, we prayed that something would change. That change came in the second demo, which Burner nailed. He did so well that the customer ended up signing. That was the second-best day in Bolt history: the day I knew our sales could scale. From that point on, with a playbook in place and proof that it could be followed, we stepped back from sales and returned our focus to recruiting.

Soon we had dozens of salespeople following the script, closing deals, and high-fiving every step of the way. We'd count how many times we could repeat our favorite Bolt catch phrases to customers —things like "it's first-grade math"—a Burner special—or "move quick or get buried by the sands of time," a Darren-ism (of which there were dozens). Darren was our second key sales hire—and he is still at Bolt four years later, as I write this book.

Seeing the benefits of hiring the right people repeated in a different setting convinced me of the universal power and importance of recruiting effectively. Going deep, building an initial playbook, and hiring against that playbook is the three-step process we used to build all of Bolt's early functions. Of course, we hired leaders who took our playbooks to the next level—and ultimately rewrote them entirely. But we learned that recruiting enables scale—leaders with great skill-sets but weak recruiting can't grow the company. Leaders with mediocre skill-sets but great recruiting do grow the company.

So…? how did we recruit world-class talent at such an early stage with such little traction?

This book distills all the learnings accrued and wisdom gathered about recruiting after four years of scaling Bolt into the company it is today. The hope is that these insights can help you in whatever stage of the hiring process you're at. As you'll see, many of the anecdotes in this book will be drawn from my specific experience building a tech company, but the lessons and takeaways should be quite universal.

FIRST: AN IMPORTANT REMINDER

There are no qualities more important in business than uniqueness and authenticity.

Anything written in this book should not be treated as gospel but rather serve as inspiration to mold your own style of recruiting.

If only a small fraction of this book is helpful, then it has served a purpose.

Okay, let's dive in.

2

FUNDAMENTALS

VISION

On day one of building your company, you may not have much—but you already have the most important thing. That thing holds enough power to pull people away from offers at the world's best companies. That thing is your vision.

Vision is by far the most lethal weapon in your recruiting arsenal. It's how you move beyond the mind and connect to the heart of whomever you're talking to. I like to recall the story of adventurer Ernest Shackleton. He recruited for his 1914 Imperial Trans-Antarctic Expedition by taking out a newspaper ad that read:

> *"MEN WANTED for hazardous journey, small wages, bitter cold, long months of complete darkness, constant danger, safe return doubtful, honor and recognition in case of success."*
>
> *Note: Please excuse the gender bias at the time!*

Thousands supposedly lined up for the job the next day.

At the earliest stages of building a company—or a team—you'll find that the people you meet are self-selecting. They will be the ones ready to embark on a journey. If they weren't ready, they wouldn't be talking to you. Now, it's your job to paint the picture of that journey—and the best way to do so is to have a vision that comes from the heart and to be honest about the perils ahead. A natural tendency is to try to hide the difficulty that will come with seeing that vision through, to make it sound easy. Here's the thing: The best people in the world don't want easy. They want challenging, fulfilling work. Uncertainty can be a draw for them. Lean into that.

I've found this lesson to be universal, even at the later stages of company-building. It's fundamental to human nature. We want to have an impact. We want to push the world forward. And, all things being equal, we will always take a job with vision where we will have impact over one without vision or the potential for impact.

Oftentimes, the only differentiator among other startups is their mission and vision. This can't be overstated enough: people join for the problem set and this is how you win talent over. Everything you do has to be centered around this to win the war for talent.

THE BOLT EXAMPLE

Our vision at Bolt is to democratize commerce and build a one-click internet.

The reason why no one has done that before is that it requires building a technology platform that can support millions of permutations of checkout and ingest thousands of unique integrations. We had to have that technology if we were going to take over the core checkout experience and not become yet another payment button. But it's really, really challenging, both in terms of the technology required and the sales effort to effectively market ourselves as a new kind of commerce platform.

The reward, on the other hand, is immense—not just financially for Bolt, but for the good of independent businesses across the world.

People understood that—and many got excited by the prospect.

YOU

On day one, you also have one other powerful ally to vision: yourself.

We often forget this point: Anyone joining your company or your team is joining because of you. Something about you energizes them and makes them feel alive. Something about you makes them trust you to lead the charge and treat them well every step of the way. Something about you gives them permission to be authentic to themselves and unleashes them in a way they haven't been unleashed before. Your personality is everything. Don't cover it up. Express it.

Authenticity is the ultimate talent magnet. This is an idea I owe to Eco's Henry Ault. Either consciously or unconsciously, we are all measuring other people's authenticity. We've trained our entire lives to suss out fakes and charlatans—and millions of years of evolution hasn't hurt, either. Authenticity communicates safety and trust. If a leader is authentic, they will attract authentic people. Lack of authenticity is dangerous. An inauthentic leader will attract inauthentic people, leading to a toxic culture.

YOUR STORY

If you're in the position to hire people—whether you're leading a team or leading a company—then you already have an interesting story, I guarantee it.

If you don't believe this, give it some thought! Your personal pitch is just as important as your company pitch. The candidate is joining for *you*, and there's no better way to deepen your bond than by sharing your story. The more you're willing to share and be vulnerable, the deeper the impression you'll leave on the candidate.

Share the hardships you've been through. Share the wins and losses. Share why you're so passionate about this team and company for which you're now recruiting. The more relatable you can be, the more candidates will feel a personal connection and be drawn to you.

You can measure success here by asking candidates why they ultimately joined you once they do. If they talk about a personal connection to your story, pitch, energy, mission/vision, you'll know you're onto something. You should do the same to anyone who declines your job offer as well—that way you have a data set to really build from and improve.

POSITIVE ENERGY

When I was interviewing candidates for vice president-level engineering roles at Bolt, I asked a simple question: "How do you recruit great people?"

One candidate was a really special leader—and they had a really special answer: "Well, positive energy can go a really long way."

It wasn't an answer I was expecting—but I saw immediately that there is so much truth in that answer. I never forgot that answer, and it has shaped my recruiting since then.

If you're disengaged in an interview, the candidate will pick up on your less-than-enthusiastic energy. It will reflect on the company—and it will potentially cost you a recruit.

The energy you exude when speaking to candidates is everything. If you are actively positive, excited, and caring, you're already in the top 1 percent of recruiters.

DON'T BE AFRAID OF BRACING AUTHENTICITY

The more authentic you become, the less you'll conform to societal standards. This is a good thing. "Real recognize real." The more real and authentic you are, the better the talent you will attract.

It's important to note that the real you can be eccentric or polarizing. This is the non-conforming you! And this too is a great thing. You might repel certain folks. This is a blessing. If you mask who you really are and pull these people in — people the real you would repel — someday down the line, it will blow up. Being real and polarizing is good: It helps filter for the energy that works well alongside yours.

Being polarizing helps with your close rate, too. If you don't stand out, you'll have a lot of good conversations, but no one will be compelled enough to take the plunge to dedicate the next several years of their *lives* to working with you.

Joining a company is an enormous commitment. Many of us spend more hours working than we do with our families. Remember this. You have to truly stand out—and it's okay to be polarizing if that represents your authentic self.

CULTURE IS EVERYTHING

In order to build a lasting company, it's all about culture. That's why at any given time 10 to 15 percent of our company is made up of PeopleOps and Talent individuals. We want Bolt to shine on the resumes of our employees, in the same way that Google did in the early 2000s or Facebook did in 2010.

That means we double down on everything from initial sourcing outreach to interviewing to onboarding and the employee journey. Ultimately, people are spending eight or more hours per day with you and your team—a substantial chunk of their time on the planet. If you can convince them that they'll have an awesome work culture and wonderful peers, that's not just making a difference in your company's future—it makes a profound difference in their lives.

This is why we took culture so seriously so early on at Bolt. It wasn't just because it was core to how we operated—it was also one of our most powerful recruiting assets. We worked on our culture daily, constantly updating and upgrading how we thought about it. Eventually, we open sourced our culture guides for the whole world to see at Conscious.org and have now turned these documents into a movement that extends far beyond Bolt. Over 90 percent of new candidates at Bolt reference our Conscious Culture movement as one of the top three reasons they decided to join our company.

We also documented our culture heavily and shared those documents with candidates. The very fact that we thought to document our culture and share it demonstrated to candidates that we were operating at a high level, relative to other companies. Very few companies or teams on the planet do this. If you do, you'll be an outlier. Don't just talk the talk, walk the walk.

PRODUCT

Use your product to your advantage. If you can get a candidate excited about your product, you've now created another differentiator and reason for a candidate to join your company.

I love taking candidates through product demos. It's a good filter, too: Do they light up when they see your product? If the answer is yes, then they'll bring that energy and passion into your office. If the answer is no, you may not want them on the team.

At Bolt we had certain individuals who would give a product demo as a part of every candidate onsite. They were specially trained on how to give the demo and make our product shine. No candidate would leave an interview without getting an exceptional product demo.

CREDIBILITY

Badges of credibility are important to candidates. They help establish trust. Don't be shy about bolstering yours in conversation. You'd be surprised at the effect this can have—and how seldom people do it.

If you have credible investors backing you, have no shame in sharing that and explaining who they are if the candidate has not heard of them. If you have a pedigreed team, take pride in walking the candidate through the team's backgrounds and achievements.

Even things like a domain name help! People who joined Bolt early blatantly admitted to me that it was largely because we owned bolt.com. Go figure!

Use whatever goods you got—and don't be afraid to flaunt those goods if they help you bring in talent.

3

FINDING TALENT

YOUR NEXT HIRES ARE A BYPRODUCT OF YOUR LAST HIRES

Silicon Valley startup wisdom suggests that your first 10 hires will dictate its next 100—and that 100 will dictate the next 1000.

There's a lot of truth to this: Hire A players and they'll attract other A players. Hire B players and they'll attract C and D players. I can't stress this enough.

If you start to compromise on the hires you make at the outset, your corner-cutting will cost you dearly in the long-run.

Why? Human beings are powerfully influenced by their sense of fairness. There are few things that will sap the motivation of a top performer than earning only as much as the weak performers.

An A player working alongside a B player and being kept at the same pay level will leave or, perhaps worse, suffer in silence. Your B players will stay and cause headaches. And the cycle will repeat as your best people leave to work with better people; more B players will then join, inviting more problems into your organization.

SAYING "NO" IS MORE IMPORTANT THAN SAYING "YES"

Because of this dynamic, the key to making great hires can be a short-term hiring drought. At Bolt, we had some hiring droughts during which we'd go six months or more without making a single hire.

There were some tough months in our Trough of Despair — a startup concept referring to the period of struggle a company faces getting off the ground. We were running out of money, customers were having untold issues, and the future looked bleak.

Even with the cards stacked against us, we never compromised on a hire. From an outside perspective, nobody wanted to join our company—but on the inside, we were the pickiest recruiters. Of course, this was a frustrating period for Bolt's managers. They needed people to do the work—and they were behind. Years later, though, they all thanked me for maintaining a high bar for talent, because ultimately they *were* able to pull great people—and those great people forever changed the trajectory of their teams.

Even when you're understaffed (or feeling understaffed), don't be afraid to say no to the wrong hire. Have an incredible filter if you want to truly build a world-class team.

SOURCING

Everyone wants to know the secrets of sourcing top talent. The good news: I'll share them. The bad news: There are few shortcuts or hacks. All paths require hard work.

If you're ready to put in this work, there are an infinite number of ways to source talent.

COLD OUTREACH

Cold outreach is the most basic form of sourcing. It's reasonably effective. Most importantly, it's reliable and repeatable.

The key to good sourcing is having a steady drip of cold outreach. In the early days of Bolt, we did this manually. I would dedicate an hour or two each day for cold outreach.

A good cold email is both an art and a science. Here's a good example:

Hey X —great to connect. CEO of Bolt here.

I just finished reading your post on ___. It's really well-written and I'm a huge ____ fan as well.

We just announced our partnership with _____ and are scaling up our team.

Are you free for a 10-minute chat and, if so, how's tomorrow at 1 p.m.?

Ryan

Notice the following things:

1. **Brevity.** Most outreach emails are way too long. They will not get read. Keep it short and sweet.

2. **No pitching.** The email doesn't explain what Bolt is. This would make the company seem small. When Google reaches out to you, they don't write "Google, powering search for the internet." They just say "Google." Leaving out a description makes you seem significant. Candidates who are intrigued and interested can then go look up the company.

3. **Proof of work.** The email demonstrates that this isn't a random or canned outreach. The fact that you've personalized it proves that time was spent learning about the person.

4. **Easy call-to-action.** Anchor on a 10-minute call and offer a time. Make it easy to say yes. Everyone has 10 minutes to spare.

5. **Not a recruiter.** A message from a non-recruiter is more meaningful, especially at the early stages with less brand power.

Channel your own tone and creativity, but keep those basic principles —brevity, confidence, curiosity, and ease—in mind throughout.

THE POWER OF A GOOD WEBSITE

At Bolt, we were stealth for many years. That being said, we had a website that was designed to intrigue. It had our values, our team, and our mission statement. It was well designed and powerful. Candidates oftentimes commented that they were intrigued by the website and therefore were open to a conversation.

TIMING

At Bolt, we found that midweek evenings—Tuesday through Thursday—after working hours have yielded the highest response rates for cold outreach messages, along with Sunday afternoon and evening. We avoid Fridays and generally suggest a time on Tuesday, Wednesday, Thursday, or Sunday afternoon for our initial conversation.

TARGETING

Define "ideal" targets. These targets' profiles should represent people that you think would be a really strong fit internally and you think might have a real interest in your company.

Sometimes, you'll start to have success in hiring from a specific company because their people are really good but morale is low. Keep sourcing from that pool if it's working. But don't go overboard, as you could be seen as overly ruthless.

As unfortunate as it sounds, in today's global talent war and with at-will employment agreements, all outreach is fair game. It is incumbent upon companies to build solid cultures, so their employees aren't compelled to pick up the phone.

LOCKING DOWN THE CALL

Make it easy for interested candidates to schedule a time to chat, either by offering several times or by including an easy-booking link (such as Calendly) for them to click and pick a time.

These emails are simple to write:

"Awesome! Can you pick a time tomorrow that works for you? https://calendly.com/your15minurl

Looking forward to it!"

IF THEY SAY "NO"

In the event that a candidate says no, sometimes this message still works:

"Understood that it may not be a fit for now. If you're open to it, though, I would still love to hop on a quick 10-min call. I've found that great things happen when smart people connect. If you're open to it, can you pick a time that works for you here: https://calendly.com/your15minurl?"

FOLLOW-UPS

Sometimes it takes four to five follow up notes before a candidate replies. These repeated follow-ups are called "Email Drips" or "Nurture Campaigns." Several follow-ups are totally fine, and often necessary.

Candidates are busy—or, more accurately, their inboxes are busy and they need reminding. The fourth and fifth email will definitely get some conversions—and it costs you little to try.

TURNING EVERYONE INTO A RECRUITER

In Bolt's early days, everyone participated in cold outreach—not just the founders. Every single person on the team committed to sending 100 messages a week.

Establish from the very outset that everyone at the company is a recruiter. This does two things: First, it spreads the burden of bringing new people in. Second, it expands the company's marketing without having to explicitly sell the product.

Think about it: Every touchpoint with your company is another way for someone to remember the name. "Oh yeah, I've heard of Bolt. Someone from there reached out to me a few weeks ago" becomes a regular refrain. On that second or third or fourth time they hear "Bolt," they may just write back—but it takes a company-wide effort to create that effect.

HIRING IN-HOUSE RECRUITERS

As soon as you can afford it, hire a full-time, in-house recruiter. This is the best thing you can do.

Talent being your most important asset as a company, having someone full-time sourcing, coordinating, and pitching candidates is enormously valuable. Recruiters spend 100 percent of their time recruiting. You can't do that yourself—so just imagine that when you hire a recruiter, your hour or two spent on recruiting is now multiplied by 8 or 10 or 12. That's extraordinary leverage.

When you start hiring recruiters, everything changes. For instance, the recruiter can create more touchpoints as the candidate's advocate internally. They can do prep calls for candidates before an on-site interview. They can do wrap-up calls with top candidates.

Companies wait too long to hire a recruiter. Or they underinvest in their talent team as the company scales. At fast-growing companies, recruiters can make up as a tenth of the workforce. I recommend hiring a recruiter as soon as your company has 12 to 15 people, assuming it is growing at a moderate clip.

HIRING OUTSIDE RECRUITERS

Some people are wary of hiring external recruiters due to cost and quality control concerns. However, if you have the budget and know how to manage them, outside recruiters can be fantastic. If someone lands you an A+ candidate, they're worth the price.

Most firms are only paid if they actually land a candidate. The exception is executive search firms, which sometimes require a guaranteed payment.

Most people will use recruiters transactionally. That's a mistake. In the beginning, I over-invest time with my recruiters. Here are some easy steps I use to build the kind of solid working relationship I need with anyone who is paid to recruit for Bolt:

1. Take your recruiter out to dinner. They so rarely get this kind of love and attention. Their motivation to help will skyrocket, as will the quality and quantity of candidates they send your way.

2. Coach them on your pitch—not just the vision, but your credentials—and yourself as a founder. Make them repeat it back to you.

3. If the second point is too time consuming, you can record your pitch and share it with them. And don't forget to share it with your team!

Try out different recruiters. If some start to work well, then stick with them. If they are not sending you candidates to your liking, tell them. Coach them on what you're looking for and build a relationship with them that can survive throughout the company's life cycle.

HIRING FRIENDS AND FAMILY

This is usually a terrible idea. Why? Friends and family are almost impossible to fire without entirely ruining the relationship. In work, the professional relationship must take precedence over the personal. When the personal relationship is too deep, many folks cannot separate the professional. There are exceptions to this rule, of course, but those exceptions are rare.

A good culture results from a hierarchy of respect and the ability to hold up certain performance standards. Mixing in family and friends can complicate things dramatically. It also impacts fairness. Friends and family may expect more of your time and attention than others do—and your other employees will see this as unfair.

CREATING A REFERRAL FLYWHEEL

As a leader, hiring friends into your team is not a great idea.

However, with regard to your teammates, the opposite is true. They know smart and capable people. And you want those people on your team!

There is no stronger way to make a hire than having a teammate recruit one of their friends. Naturally, people trust their friends. If you respect your friend and they just joined a company and they're telling you to join too, that's about as powerful an endorsement as there is.

The best way to inspire this, of course, is to have a great company, traction, and culture.

Even so, there are ways to boost your friend-of-an-employee referrals, including:

INTERNAL REFERRAL PROGRAMS

Paying for referrals can encourage them, of course, but be careful about creating an overly transactional situation.

Early on, instead of paying cash for referrals, we took the teammate who made the referral (and, optionally, their spouse or partner) out to a Michelin-starred restaurant if their referral was successful. This was a way to pay with something much more meaningful than cash: your time.

Obviously, this approach doesn't scale, but in Bolt's early days, it worked like a charm. Everyone wanted a fancy dinner with the founder.

Instead of cash, stock is a great option. This sets up a culture of missionaries instead of mercenaries. People sense a link between the quality of a company's people and the potential of its equity. That's powerful—and it keeps people thinking about the success of the company in the long term.

In addition to setting up the right program, the company must beat the drum. Create a fun leaderboard. Talk about recruiting constantly. Without keeping it top of mind, it's easy to forget—like anything else, recruiting can fall off the to-do list if you don't discuss it regularly. We:

- Talk about hiring at town halls,
- Celebrate who did the most interviews,
- Announce all new hires at team get-togethers,
- Hype folks up internally (i.e. via Slack) when they get hired, and
- Gamify sourcing.

Everything we do is centered around recruiting, so much that it's the heartbeat of our company. We created a "culture of recruiting."

RECRUITATHONS

Sometimes, you need everyone to sit down, comb through their network, and begin outreach.

The best way to do this is a Recruitathon. It's just like a Hackathon, but for recruiting! Frankly, I'm surprised this isn't a more standard practice.

Find ways to make it fun. If you're in an office, buy food. If you're remote, give everyone a meal credit and hop on a Zoom call.

Start with some training, tips, and tricks. Then, let the games begin!

IN-NETWORK RECRUITING

"Your network is your net worth"—this is another favorite saying from Bolt's early days.

Your network is one of the most effective ways to source candidates. My all-time favorite computer science professor, Jerry Cain, used to work at Facebook. Eric, my early cofounder at Bolt, had the idea to reach out to Jerry for referrals. I was in total shock that Jerry agreed; he made an introduction with an old coworker of his. We ended up hitting it off and hiring the connection, an engineer leagues above anyone we had hired before.

Understandably, not everyone knows Stanford computer science professors who were also early Facebook employees. That being said, everyone has a network, no matter how large or small. Ask everyone and anyone you know if they know people who would be good fits for the roles you're looking to hire. You never know who will turn up.

EVENTS

In Bolt's early days, we hosted Sunday Jam Sessions where we invited tons of founders and engineers from our network. We had dubstep playing. Super Smash Bros was loaded up. We offered tons of good food. About 25 percent of the time, people were working; 75 percent of the time, people were just hanging out. It self-sorted for some pretty awesome people. These Sunday Jam Sessions were absolute magic.

Later, we started doing game nights. We had 50-plus different games available, from board games to video games. We brought food and drinks. These game nights were epic, too.

I could not recommend these casual events more highly. Even if they're not your style, do events that are congruent. It's WAY easier to invite someone to an event than to directly recruit them to your company.

You can begin to intertwine this into your cold outreach as well.

Instead of the call-to-action about a role at the company, invite them to something fun. Fun often sells better than formal recruitment.

SOCIAL MEDIA

I really didn't start using LinkedIn or Twitter properly until we were a multi-billion dollar company. I took a lot of pride in staying "low-key"—and there were benefits to this approach.

There are pros and cons to both remaining in stealth mode and to being public-facing. Being stealth keeps your team focused and keeps you off your competitors' radars. The latter factor was very important for us at Bolt. We knew we had a long way to go before our company could scale. If other large companies found out about what we were doing, they would copy it before we had enough scale to be competitive.

We became more public-facing when we reached "escape velocity": When we were moving so fast that it would be difficult for others to catch up. And when we did become more public-facing, copycats popped up and big companies gave them tons of capital to compete with us. Thankfully, we were far ahead by that point.

Stealth also adds some mystique when recruiting. We had a mysterious website around the future of payments on a killer domain, bolt.com. This piqued peoples' interest—everyone wants to know what's going on behind the velvet rope.

When done right, being highly visible has tremendous benefits for recruiting, sales, and more. But, it comes at the cost of alerting your competition to what you're doing.

When I finally started leaning into social media, I was blown away by its impact on recruiting. With minimal work, my LinkedIn posts would generate thousands of views—and soon tens and hundreds of thousands. I then got better at leaning into "my voice," which led me to Twitter and a large audience there. My tweets started getting millions of views. My follower count grew. I believe this was in large part because I was credible and because I had built a really successful company before I began posting. I walked the walk before I talked the talk.

Every single one of those social media views helps people keep me and Bolt at the top of their minds. When someone sees you pop up once, it's fleeting. When they keep seeing you pop up with news or interesting thoughts, they're hooked. Recruiting is all about staying on the radar.

In addition to building your own presence, your team should be involved. In fact, a social media post from a teammate will likely have an even higher impact for recruiting than one from a founder. We repeatedly encourage the team to post on social media and give them facts, figures, and even custom-made images and videos that they can share.

CAREER FAIRS

Career fairs are excellent for recruiting junior-level talent. Make sure to train your staff on how to use them effectively. Train your team to talk about their direct experience at your company. This is the most powerful selling tool. Train them on how to gauge interest. Perhaps they are collecting emails and reaching out afterwards. Maybe they're scheduling intro calls on the spot. Train your team to be high-energy and encourage them to communicate the vision of the company with enthusiasm.

Here's a sample before-event, during-event, and post-event checklist we used:

BEFORE EVENT

- Create Facebook event (clone the prior event)
- Get everyone on the team to invite people in their network
- Send invite one to two weeks before the event
- Ping one week before the event
- Remind them again the day of the event (last push to increase turnout)
- Build up around the reveal to make people feel like that can't miss the event
- Buy more material than you need
- Think through how your setup can make it easy for lots of people to come (play area for children, free Ubers home for employees)

DURING EVENT

- Focus on guests, ensure they feel welcome
- Introduce guests to relevant team members if it seems appropriate or you find commonality in background
- Have fun! People remember how you make them feel more than what you do or say. If you have a positive vibe, this impression will last and influence future engagement with Bolt
- Have someone take a photo of the group
- Post once in the Facebook event to encourage turnout

AFTER EVENT

- Next day: post another photo thanking everyone for showing up
    - This serves to make those who showed up feel like they were part of something cool. It also shows those who are on the invite list but didn't show that they missed out on a good time and that they should want to come to the next event
- Ping ALL your invitees after the event (regardless of whether they came to the event)
    - Even if they didn't show up, you now have a dialogue open to invite them out again

The funny thing about checklists is that, for such a simple tool, they are remarkably powerful. Even just making sure that these basic things are completed puts you ahead of the pack.

ONLINE JOB BOARDS

There are a lot of online tools for posting jobs. Don't hesitate to use those either. Talented people can come from anywhere. Sometimes they have college degrees; sometimes they don't. Sometimes they have work experience; sometimes they don't. Sometimes they live in places that have career fairs; sometimes they don't. By putting your jobs up online, you might just find that candidate who lives in a place you'd never visit—but has the exact tools, drive, character, and motivation that you're looking for.

Think of job boards a bit like creating a few more ponds to fish in: You don't have to do much to suddenly increase the number of prospects.

4

A VALUES-DRIVEN CULTURE

Remember: You aren't just hiring or recruiting one person, you're building a culture. This means you're also assessing each potential hire at a value level to see how they'll fit with your team and mission. We have criteria built into our interview scoring guides to make sure we're testing for values alignment:

- **Passion:** Does the candidate get visibly energized when talking about past projects or topics that interest them? Are they motivated to make an impact? Do they have a hunger to learn, grow, and accomplish? Do they have fire? Are they deeply curious? Do they get obsessed with the things they do?

- **Ownership:** Does the candidate take on things outside of their job description? When they identify problems, do they also identify solutions? Do they fix those things themselves when possible? Can they go really deep when talking about past projects?

- **Scrappiness:** Can the candidate move fast with little direction? Can they make decisions without all the data and resources that a large company has access to? Are they an unrelenting problem-solver? Do they come up with creative solutions to problems that at first seem impossible?

- **Collaboration:** Does the candidate ask follow-up questions? Are they unafraid to offer their opinion? Do they listen carefully to your thoughts and ideas—or are they just interested in advancing their own? Is the person upstanding—in other words, a mensch? Would they go out of their way to support others? Are they a team player? Do they work well with others and lift up the whole team? Are they respectful, mature, and thoughtful?

- **Empathy and Inclusivity:** Does the candidate do things without expectation of anything in return? Do they appreciate different perspectives? Are they aware of their own weaknesses? Do they make others feel comfortable around them? Can they put themselves in others' shoes?

- **Coachability:** Does the candidate take feedback well? Do they ask for feedback? Did they make changes based on past mistakes? Can they both give candid feedback in a respectful way and embrace the feedback of others?

It's unlikely to answer all of these questions—but it is possible to get a powerful directional sense, especially when multiple people on the team are interviewing someone and offering their insights. Test for these value indicators — they're often a more powerful tell about a candidate's potential than whether they have the skills on paper to succeed at the job.

Part of working at a start-up is having to level-up skills when new and unforeseen challenges arise. Qualities like "ownership" and "scrappiness" are critical in that kind of fast-paced environment because when some fresh problem arrives, great teammates are able to own it.

HOW TO INTERVIEW FOR VALUES ALIGNMENT

To test for those ingredients, each interviewer should ask one or two questions assessing a candidate's fit with the company's culture and values. Here are some great questions to try:

- What motivates you?

- Tell me about a time that you didn't know how to do something. What were the challenges you faced? What was the outcome?

- Can you tell me about a time when you didn't have adequate resources for a project but you figured out a solution anyway?

- How do you incorporate giving and receiving feedback into your daily work life? Can you give me an example?

- Can you tell me about a time that you had a conflict or disagreement with someone professionally? How did you resolve or learn from the situation?

- How will this role challenge you?

- Which of our company's core values do you most identify with? Which do you least identify with?

- What's a past project that you're proud of? (See how passionate they get when answering, then ask followup questions.)
    - What part of the project did you do? What did the team around you do? (Make sure they are not taking credit for work their team did, as opposed to their own contributions.)
    - What was the most challenging part of the project?
    - What was the toughest decision you had to make? Why did you make that decision? What were the alternatives and tradeoffs?
    - What were you able to deliver, and in what timeframe?
    - Did anything unexpected come up? How did you deal with it?
    - What mistakes did you make? What did you learn from them?

Adapt these questions for your own company's needs, but don't forget the guiding principle: You are testing for values as much as you are testing for competence.

"CULTURE ADD" VS. "CULTURE FIT"

One of our core Bolt values is "United and Unique." We like hiring unique individuals who stand out from the crowd. Instead of hiring for culture fit, we hire for "culture add." To assess that, we ask the following questions of ourselves when we make hiring decisions:

- What unique quality does this candidate bring to your team?

- What about them is going to push us and our team to think and act differently?

- Do they have a special sparkle about them?

- What are the person's values? Are they fundamentally consistent with ours?

- Will this person augment our culture? Will they fill in a gap that exists? Will their perspective enhance our ability to solve problems or be an effective team?

A major pitfall of hiring is that you will naturally get excited about someone who already fits your team's culture. Be careful with this: While candidates should share the same values as your team, they should not get favorable treatment for looking the same, acting the same, or having the same personality.

"Culture add" does not mean asking yourself "do I like this person?" or "if I was stuck at an airport, would I enjoy chatting with this person?" Questions like these too often boil down to "is this person like me?" That creates a monochromatic culture—and leads to blindspots in your team, your product, and your company.

"Culture add" also does not mean "does this person clearly check every single one of our values?" That expectation is unreasonable, virtually un-assessable in an interview, and unnecessary. We've heard people say, "this person was quiet, I'm not sure they are a culture add." Often, quiet people are some of the most powerful teammates.

So… go ahead and hire that person who's timid, thoughtful, and reflective. They could change everything.

Look out for any red flags—evidence that the person may operate against one or more of the company's values—but also search for people who will stretch your organization, make you think differently, and push you to be better. If given a choice between a candidate who is a "culture fit" and feels like a carbon copy of another teammate and a candidate who is a "culture add" who could stretch the company, choose the person who will be additive, not repetitive.

EVERY HIRE MUST HAVE A "STRONG YES" CHAMPION

Every person hired should have at least one champion: an employee who believes so strongly in a candidate that they're willing to say that they are a "must-hire."

The champion must see something extraordinary in the candidate, something that makes them believe the candidate will not only be a strong contributor, but level-up the team around them and bring 10X ideas to the fold.

This does *not* mean:

- **The candidate is "perfect"**—the candidate could be someone who has mixed scores, but for some reason the interviewer sees something special.

- **The candidate is charismatic**—this is a common pitfall. Purely being social should not make someone a "must hire."

- **The candidate fits in**—this is another common pitfall. Someone who fits in with your team or culture isn't adding anything. In fact, creating a monoculture is generally a bad thing.

This is important because it adds weight to the hiring decision and forces everyone to commit 100 percent to every person who is extended an offer. If an interview panel has no champions—even if everyone is a moderate yes vote— pass.

DEI

To build a lasting company with a thriving culture, it's essential to focus on DEI (Diversity, Equity, and Inclusion). Greatness comes from having a diversity of voices, backgrounds, and perspectives, which ultimately lead to the best ideas. So if your organization is striving to do something great, diversity is fundamental.

Unfortunately, the tech industry has a pretty dismal record of attracting and retaining talent from underrepresented groups. And those groups are often treated unfairly across the employee lifecycle, from hiring to pay to promotions. This needs to change.

Diversity means differences in gender, gender identity, race, ethnicity, sexual orientation, age and socioeconomic class in the workplace. Diversity breeds more diversity. The longer it takes a company to focus on diversity, the deeper of a hole it has to climb out of. If your company is 90 percent male, and a female candidate comes to interview, how do you think she'll feel when everyone interviewing her is male?

Equity means fairness in the workplace. It's about recognizing the different circumstances of people and systemic issues that could mean treating people differently in order to create a level playing field for all.

Inclusion means ensuring everyone feels a sense of belonging in the workplace. What is your company doing to make sure all employees feel comfortable bringing their full authentic selves to work? This may look like creating affinity groups, educational events, a budget for employee development and more.

Great candidates know that great companies value DEI and will be testing for it during the interview process. They'll be on the lookout for a diverse panel of interviewers, a sense of fairness in the organization, and a feeling of comfort and authenticity.

Too many founders only focus on DEI after the company's dearth of diversity becomes a problem. But once you have a diversity deficit, it's hard to rebound.

Don't wait. DEI needs to be priority one, day one.

5

THE PROCESS

Let's say you've sourced a candidate, and they've agreed to talk. Now what?

INTRO MEETING

This is your first conversation with a candidate. This call is 75 percent selling and 25 percent interviewing. At this stage, you're asking some basic interview questions, outlined later in the book.

Whenever possible, do your initial screens in-person or on video. On average, 30 to 40 percent of your first video screens should be making it through to a second round, but you should not be pushing every available candidate forward.

Here's the information you're trying to gather from this conversation:

- What is important to the candidate?

- What are their motivating factors for changing jobs?

- What are they looking to gain from their next role?

- Fill in any missing information about their experience.

- If you sense that they're going to pass the first screen:
  - Make them feel important
  - Sell the candidate on the onsite recruiting experience
  - Plan next steps

For a candidate to move forward, it should be clear that they have the required skill-set, are a culture match, and fall within the role's compensation, job level, and reporting expectations.

With only a minority of candidates making it through initial video screens, you can build deep relationships and double down on your connection with advancing candidates. You must demonstrate deep candidate empathy in order to establish the candidate control you'll need to close.

Your other job is to make sure that the candidate is intrigued enough by the end of the call to commit to next steps.

As a way of closing out the first call, here's a recommended script:

"I'll take the notes from this call and details from your profile and share them with the hiring manager. We should be able to get back to you in a day or two."

It is crucial to stick to whatever time commitment you give the candidate for following up. Each part of this process is about building a relationship based in trust and honesty; timely follow-ups help you do that.

DEEP DIVE

This is a longer conversation that's 50 percent selling, 50 percent interviewing.

I tell candidates to come prepared with a bunch of questions—and I say that I'll be doing the same for them. The point of this conversation is to really go deep, to answer their questions and see how they answer your own. It's to assess what it would be like if they were whiteboarding with you—how might it feel to problem-solve with this candidate in real life?

At the end of this process, you should know how seriously you're interested in the candidate and vice versa. If it's going in the right direction, pitch a full on-site interview (either in-person or remote, if necessary).

I pitch the on-site by saying something like:

"Well, I've really enjoyed this conversation. If you have too, I can suggest next steps."

"Sure."

"For our next step, we'd like you to meet a few more folks on our team. This will be dual purpose: These will be interviews, but also an opportunity for you to meet more of the team and get a bunch of your questions answered. Whaddaya say?"

In the early days before talent demand was in Bolt's favor, I'd like to get the candidate committed to a time before we hung up the call, but that's not always necessary.

FULL ON-SITE

What most companies forget about the on-site experience is that it exists as much for the candidate as for the company. Great people will be evaluating you and your company during this experience. It's your job to blow them away.

Make sure the on-site is neither too long nor too short. I recommend at least three full-length interviews and no more than five. Make sure that junior teammates aren't interviewing candidates for more senior roles. This is usually a disaster because, by definition, they aren't senior enough to evaluate for that role. If you do have junior teammates involved, position it as a meet-and greet, not an interview.

Make sure the candidate is greeted, made comfortable, offered water, and shown around. They should ideally have an on-site buddy who is guiding them through the entire process.

Interviews should be highly collaborative. Avoid "gotchas" and questions to which there are "right" or "wrong" answers. Give candidates hints when they need them.

If someone prefers a whiteboard over a keyboard, cater to their preference.

Make sure not to ask the same questions over and over again. Be coordinated so that all interviewers are asking a different style of question.

If you can, give the candidate a parting gift—a company t-shirt works well. The small things matter.

DEBRIEF

All on-site interviews should be followed by a debrief with the interview panel, even if the decision feels obvious. The goal of the debrief is to evaluate the candidate based on everyone's shared feedback, identify possible improvements to the interview process, make a decision on the candidate, and discuss next steps, if applicable.

The process can flow as follows:

1. Schedule the debrief, ideally at the end of the day or within 24 hours of the on-site interviews.

2. Everyone involved should write feedback beforehand, based on each person's assigned evaluation role.

3. At the assigned time, everyone who interviewed the candidate and the recruiter should gather, almost always in a conference room or private area outside of general earshot.

4. The recruiter then reminds the team of both the role and what team members interviewed the candidate about what competencies.

5. The recruiter should ask each participant to share their feedback about the competency for which they interviewed, keeping the conversation limited to the set of things a person was evaluating against or that are within their area of expertise.

To maximize debriefs, I've found the following helpful:

- Each person gets under two minutes to speak. This ensures that your debrief doesn't turn into just another off-site interview that chews up everyone's day.

- Make sure to touch on the candidate's grit, authenticity, skill-set, menschiness, and other qualities.

- After each person has shared their feedback, we give them 90 seconds to make their case—to hire or not to hire.

- The hiring manager has the final say, subject to veto from the "Bolt Calibrator"—a name we give to the person in our organization responsible for keeping the bar for hiring consistently high. If a hiring manager feels very strongly that the Calibrator is wrong, we can set up a follow-up conversation with another person responsible for bar-raising.

This is how the process works at Bolt, but the key thing to take away is that we have specific protocols not just for talking *to* candidates, but *about* them after the interview. This is how you leverage the collective intelligence of your team to ensure that new people are a fit—when it comes to hiring, two heads are definitely better than one.

MULTIPLE TOUCH POINTS AND MAINTAINING MOMENTUM

The more high-quality touch points you have with a candidate, the better. When dating, if you don't follow up after a date, you're probably not getting very far.

And as with dating, momentum is key. Make sure next steps are laid out, and new touch points created—especially for candidates who are in high demand. In addition to following the process outlined here, it never hurts to ask a candidate to hop on a 10-

minute call to discuss where they are in their job search process, and it never hurts to email the candidate with some new article you think they might find interesting. The holy grail is to be on a texting basis with candidates. Don't overwhelm them, but if you're texting, you're in a really good spot.

Don't be annoying, pushy, or rude. Use tact and restraint—and make sure your touchpoints are genuine and sincere.

REFERENCE CHECKS

Over the lifetime of our company, Bolt has done over 10,000 reference checks. Crazy, right?

Not really. We've found them to be among the most powerful tools in our talent arsenal. References important for three reasons:

1. **Evaluation:** References provide us with critical information

2. **Alignment:** References create a tighter hiring process

3. **Communication:** The process of reaching out to references gets the candidate excited about the company

Number three may be counterintuitive. Here's what I mean: The depth of a reference call makes all the difference. A simple checklist of questions is annoying, bland, and cumbersome. Thoughtful two-way questions are essential. Candidates step back and say, "Wow. If the company cares about me this much now, imagine when I'm there." This is big.

One important note: For every hire, ask for three references. The first should be someone they reported to. The second should be a peer. The third should be someone they managed. This will give you a 360-degree view of the candidate.

Here are some questions to use:

QUESTION 1:

What is your professional relationship with the candidate?

Also ask what the candidate has told the reference about Bolt and the position at the company. This will allow you to give any necessary background information about your company and the job you're hiring for.

This question breaks the ice and subtly gets the referencer excited about your company. Here's a secret: If this person is a reference for someone you want to hire, chances are good that they're likely a strong candidate in their own right! This can be helpful down the line. Additionally, the reference will likely speak to the candidate after your call. Crushing the casual pitch can help you nail two birds with one stone.

QUESTION 2:

Tell me what it's like to work with the candidate.

You can get a lot of information from an open-ended question like this. Instead of diving right into specific questions, stay vague. Let the reference speak in order to get as much information as you can.

QUESTIONS 3/4:

Over the course of your career, how many people with this role have you worked with who have comparable experience? How would you rank the candidate among those people?

Try to get specific numbers. Another secret: I've never heard someone describe a candidate as "bottom 50 percent." Everyone is always "at the top." Mathematically, that obviously can't be true. Dig in to figure out how "top" they are—are we talking top 50 percent, top 10 percent, or top 1 percent? There's a big difference between those answers.

QUESTIONS 5/6:

What makes them one of the top 10 percent of people you've seen in this role? What separates others in the top 5 percent from this candidate who's in the top 10 percent?

This is another way of asking about strengths and weaknesses, but with real data points. It will lead to better answers.

QUESTION 7:

If I were reading this candidate's peer reviews, what is an area of improvement that I might uncover?

This is yet another way to ask about possible areas of improvement. It's safer to ask it this way because asking about peers allows the reference to put their own identity and thoughts aside.

QUESTIONS 8-12:

How well did the candidate get along with their co-workers and management?

What kind of personalities did they work well with?

On a scale from one to five, how coachable are they? Why did you give them that score?

While you worked with them, where did you see the most growth?

These are pretty straightforward but important questions. Coachability and ability to work well with others are both critical to any successful teammate—particularly managers. Always make sure to ask about these essential traits.

QUESTION 13:

If the candidate were to join us, how would we best set them up for success in the first 90 days?

If you're doing references on a candidate, you might as well prepare yourself for what comes next: Managing the person if they are hired. If you're looking to get advice on how to manage and onboard someone new, the best person to ask is someone who has just worked with them—and that's who you're on a call with! This question ends up being one of the most valuable we ask.

QUESTION 14:

What is something you haven't told me that you think we should know? Or, alternatively: What is a question that you

think I should ask that I haven't asked?
I like to start reference checks with an open-ended question—and I like to end with an open-ended question.

If you do reference checks this way, they can be incredibly powerful. The person you spoke with will be blown away by your level of thoughtfulness. You will also come away with a much more accurate picture of the candidate you're considering.

HOW TO REACH OUT TO REFERENCES

Here's a really simple email and text template for how to reach out to references:

Email

Subject line: Reference for [CANDIDATE FIRST NAME]
Hey [REFERENCE FIRST NAME],

Great to meet you via email. We're in discussions with [CANDIDATE FIRST NAME] about an opportunity at Bolt and they said you would be a good person to talk to as a reference.

Do you have time for a quick (~10 min) call any time over the next couple days? Please let me know when is most convenient for you. Alternatively, feel free to call me at your convenience at [PHONE NUMBER].

Thanks!

Text

Hi NAME – [CANDIDATE NAME] suggest I reach out for a reference. Do you have a few minutes to connect? – [YOUR NAME]

REJECTING CANDIDATES

Every now and again, you'll go through the initial screening process, the deep-dive interview, the on-site experience, and do a thorough reference check—and then realize you don't want to move forward.

That's when you have to reject a candidate. This isn't easy—at least, it's not easy if you care about people and don't want to disappoint them. I know it wasn't easy for me the first few times. But along the way, I picked up some tips to reject people with empathy, honesty, and decency. Here are some things I've learned:

- Always schedule rejections for one to two days from the on-site interview—do not send the message immediately, even if you came to a decision immediately. The deliberation gives you the chance to communicate better and gives the candidate a better experience.

- Try to send rejections on weekdays, but avoid Friday. Schedule the email to go out at the end of the day on Monday if it is Friday when you decide to reject the candidate. You don't want to ruin someone's weekend.

- When possible, provide candid feedback. This would make the most sense if the candidate made it to an on-site interview. Try to keep the focus clear and simple. For instance, I might write: "You did a good job in the interviews, but fell short in the presentation portion." You'd be surprised what can happen when you give someone this feedback—sometimes it's the very thing they need to ace their next interview. This is mostly relevant for senior candidates and referrals who will be comfortable with and accustomed to this kind of feedback. Interestingly, how they respond to the feedback might tell you a lot—it's just an additional data point in a relationship, which is what recruiting is ultimately all about.

As with all parts of this process, you should conduct rejections with humanity and care—and if you do, you'll send a powerful signal about your company and its values. Internally, people will also see that even if someone doesn't make the cut, they are still treated with grace. That sets a tone that can echo powerfully throughout an organization.

6

CLOSING & THE OFFER

Closing depends on having done everything well during the interview process. By the time you're ready to close, the candidate should have a strong sense of your culture, feel like they met a high-quality cast of teammates, and noticed both the speed and thoughtfulness of your interview process. They should feel like a special relationship has been forming through the multitude of touchpoints. And they should believe some combination of your vision, product, mission, team, or culture stands apart from the other companies they're considering.

Remember, there are two ways a candidate makes a decision: with their heart and with their mind. Ideally, you strive to win over both. Their rational mind must be convinced that this is a strong decision for their career and their family. Their heart should feel a special connection with your team and your company.

In order to close a candidate, work through both the rational and the emotional.

THE RATIONAL

EQUITY MECHANICS

If equity is part of your offer, it's a great service to be able to take the candidate through how to think about equity. This builds trust and helps the candidate understand what is likely the most important component of their offer.

The goal of this conversation is to help the candidate understand the value of the equity today and tomorrow. If your startup is in a very early stage, the value today might be quite low, so the emphasis must be on the value tomorrow.

Assume the candidate knows nothing. Take them through the basics, including what a strike price is. Here's a quick 101 of the basics an equity offer:

- **Strike price of equity:** This is the 409A price that stock can be purchased for. If the 409A strike price is $0.05 per share and the offer is for 1,000 shares, then the cost to purchase the shares is $50.

- **Dollar value of equity according to investors:** Typically, the price investors are willing to pay for stock is much higher than the 409A price. Investors are usually purchasing Preferred Stock, which has a liquidation preference. A liquidation preference means that, upon a liquidity event such as an acquisition, their money gets returned to them first before the proceeds are distributed among the cap table. This is only for an acquisition; if the company makes it to IPO this right gets removed. Employees, on the other hand, receive common stock, which does not have a liquidation preference. Usually, the investor value is used to show the market price of the equity. So, if the investor price is $0.20 per share and the offer is 1,000 shares, then the investor value of the shares is $200. This $200 figure is typically used to characterize the size of the equity offer.

- **Restricted stock versus options:** Restricted stock is purchased immediately at the strike price and the employee owns the stock from the outset. Options allow the candidate to purchase the stock at the locked-in strike price today, but also at a point in the future. Candidates usually have 90 days from termination to exercise their options, although some companies are extending exercise windows to years from the date of termination.

- **A special technique:** There is something that startups typically allow for their key executive hires: early exercise. This allows the hire to exercise early, which starts the capital gains clock. In addition, the company may give out a long term (i.e. five-year) loan for the exercise price. This is a cashless give because the money is then paid back to the company to purchase the stock in turn. So, there is no cost to the company, and immense benefit to the employee. Traditionally only done for executives, Bolt decided to break the norm and offer this benefit to every single employee. There is some risk to the employee if the stock goes down as they'll still be on the line for the loan. That being said, we paid for them to get third-party financial advising in order to make an educated decision. Many employees took us up on this offer and were immensely grateful.

EQUITY POSSIBILITIES AND EXPECTED VALUE

Expected value is a powerful concept for thinking through life decisions. The expected value is the anticipated value of an investment at some point in the future. Joining a high-growth company is all about expected value.

The way to calculate expected value is by multiplying the possible outcomes of an investment by their expected probabilities.

This chart is designed to lay out expected value for candidates you're in discussion with. Let them fill in the "Probability" column with their own percentages. This screenshot is a real Bolt offer made several years ago. This $120,000 four-year equity offer, granted when Bolt was a $200 million company, would be worth over $3 million as of November 2021.

Salary	Shares*	Valuation today	Equity Grant	Dilution Factor	Curr Annual Equity Value	Total value
$115,000	6,131	$200,000,000	0.08%	20.00%	$30,001	$120,006

Hypothetical Valuations	4-year equity value	Annualized	Probability	Annualized x Probability	Total 4-Year Expected Value
$0	$0	$0	30.00%	$0	$0
$200,000,000	$120,006	$30,001	30.00%	$9,000	$36,002
$300,000,000	$180,009	$45,002	15.00%	$6,750	$27,001
$500,000,000	$300,015	$75,004	8.00%	$6,000	$24,001
$1,000,000,000	$600,030	$150,007	6.00%	$9,000	$36,002
$5,000,000,000	$3,000,149	$750,037	5.00%	$37,502	$150,007
$10,000,000,000	$6,000,298	$1,500,074	4.00%	$60,003	$240,012
$100,000,000,00	$60,002,978	$15,000,744	2.00%	$300,015	$1,200,060
		Annual expected value:		**$428,271**	**$1,713,085**
		Expected value over 4 yrs:		**$2,173,085**	

*Equity is granted as ISO

The problem with using these hypothetical valuations is that they are tough for a candidate to believe. Bolt has already surpassed many of the valuations listed, but at the time, it was difficult to convince candidates that we would.

That's why it's important to reference the valuations of adjacent or similar companies. This anchors the numbers, and can make them seem more achievable. For a good example of this, refer to the Masterpiece Offer Letter section below.

The small percentages tied to the outsized hypothetical valuations are what really move the needle. If you're building something world-changing, candidates should believe there's at least a few percentage point chance that you build a $100 billion company. If they don't believe that, then they might not believe enough in where you're heading.

THE EMOTIONAL

CLOSING DINNERS

Closing dinners crush it. If you can get the candidate's significant other to join, even better. Remember, spouses and significant others are key decision-makers!

WELCOME BOXES

Another nice touch is sending welcome boxes to a candidate's house. Include items for kids if they have any. Win over the entire family.

TEAM CONGRATULATORY NOTE

Once we give a candidate an offer, we have everybody they spoke with at Bolt reach out to them. The benefits are twofold: The note is a friendly touchpoint that reminds candidates of the team they could be a part of and it also gives them an easy way to maintain a productive dialogue as they weigh their decision. Many will circle back to ask specific questions of various potential teammates.

At Bolt, we are deliberate about creating a culture around

inclusivity and enthusiasm. After the candidate has said yes, we often have their hiring manager or the lead for the team make a social post—as long as the new team member is okay with it—on why the hiring team is so excited this person is joining Bolt. A good LinkedIn post on someone joining can generate 100s of reactions and 1000s of views—and it can be the kind of thing that makes the candidate feel incredible about joining your team.

PRESENTING AN OFFER

How you present your offer can matter as much as the offer itself. Here are some steps to avoid delays—and the possibility of losing a candidate because of them:

1. Schedule a debrief for the same day or the next business day after the interview.

2. If the candidate is targeting a salary well outside of your salary range, bring that fact up early with your team—even before the candidate comes on-site—so that everyone is aware of this and knows what needs to happen if we end up wanting to move forward.

3. Proactively ask about the candidate's availability for any follow-up conversations you may need to have.

Once you've decided that you're moving forward with an offer:

1. It should never take more than 24 hours to get numbers approved and the potential offer sent to the candidate.

2. You need to make sure there are no delays in this process—keep positive momentum with candidates. Candidates on a tight timeline need to be given numbers as soon as possible. Remember that you might be competing against other companies. Time is of the essence.

3. Confirm a candidate's interest before presenting them with an offer. Typically, you can find out by asking them if they can see themselves working at your company or how they're comparing your company to others. Answers to these questions are imperative to know if you decide to present an offer and can

give you information on how aggressive you need to be with your offer.

4. Consider whether it's better to wait on making an "official" offer. If the candidate is hearing about other offers from companies within one to three days of hearing from your company or if your company's offer will be the last they receive, it can be beneficial to give an informal offer before you have your official offer ready.

5. It's best to present an offer when the candidate is most excited about your company—and often this means after speaking with the senior-most leaders. So wait until those discussions happen and then present the offer.

If a candidate says they need an extended period of time to make a decision, you should evaluate if they are serious about joining you. If you've done everything mentioned in the previous pages, it should be rare that a candidate sits on an offer from you only to come back weeks later and accept. In that instance, it might be best to keep them warm—through conversations, further interviews, dinner, and regular communication—and continuously touch base to gauge their interest and find out what needs to be done before closing.

THE BIG EMAIL TECHNIQUE

This is one of my all-time favorite recruiting techniques.

Sometimes, there are a lot of reasons that someone should join your company, but for some reason that candidate can't seem to hold them all in their head.

In these cases, I like to send an email — a long one. I outline my entire position on why I believe they should join the company. Of course, this is only if I genuinely believe that they should join us. If I don't believe they should, I move on.

Most importantly, I use their words. Right at the top of the email, I list out everything that I heard that mattered to them during the interview process, in as much detail as possible.

This shows you listened. I learned how critical this was through the Mochary Method, the brainchild of Matt Mochary, one of the best mentors I've had. Whenever you feel like you're in between steps and the candidate needs an extra dose of energy, this email can work phenomenally well.

Here are some real emails I've written to candidates close to making their decision. This first email was to a major executive who ended up on our team:

[NAME],

To say I'm excited about the possibility of working together is a wild understatement.

Not to get too mushy, but it feels like we've known each other for a very long time. I have met thousands of folks on this Bolt journey, and to meet someone with this level of resonance doesn't really happen. In my eyes, there is no doubt that the universe is willing a divine partnership into existence.

On that note, I wanted to share some thoughts over email. Many of these points may seem redundant, but I am always compelled to get important things in writing →

1. No one will go for it like we will. Every step we've taken—from setting culture right to scalable technical architecture—has been with this long-term vision in mind. We've ensured that nothing will get in the way of building the biggest business imaginable. More important than these logistical limitations are the limitations in our own minds. Together, we will think bigger than anyone.

2. No one will have more fun than we will. Every encounter thus far has been so much fun. I had a very similar experience with _____ during our interview process. This did not stop after we got to work; in fact, the opposite is true: Our trust deepened, our achievements compounded, and our level of fun has only risen with every passing week. The more we win, the more fun we will have along the way.

3. We will fundamentally change payments and commerce. There will be no more important company than Bolt in the history of this space. In three years' time, we will be bigger than X, Y, and Z. Our business model is powerful beyond belief. And our network-driven strategy hits the weak spot of all the major incumbents. We will write history.

As far as economics go, we're willing to extend an offer to you at a present-day value of roughly $__ million in equity.

We are all gunning to build a $100 billion company in the next three years. Excluding some dilution, this would make the equity worth close to $__ million. Even at this stage, we will still just be getting started.

As far as cash goes—you tell me what you think is generous but fair, and we'll approve that number. Equity is what matters most here, and on the cash side we want to make sure you're compensated at a level you're comfortable with.

I really hope we will work together, but I respect whatever decision you ultimately make.

Ryan

This email was to another major executive who also ended up joining Bolt:

Hey _______,

Thank you for the last couple conversations. I agree that before diving into final steps and meeting the rest of the team, let's try to nail down ballpark economics.

Several things are clear to me:

1. You're motivated by ownership in the company more than cash. If you're going to invest the next five to 10 years into Bolt, you don't want to have to think twice about your ownership.

2. Upside is great, but you also have upside at ____, and you also want to be clear on the present-day value of whatever offer we put forward.

3. You are not only evaluating this opportunity against your situation today, but also what you could do with your career going forward (higher ranks at current company, running another company as CEO, etc.).

4. You are exceptionally talented and a league above the others we're considering for this role.

With that, we wanted to be extra thoughtful here and investigated across several vectors. Most importantly, we want to recognize the exceptional level of experience that you can bring to Bolt. While our expected offer range was between X-Y percent, and that assumed an exceptional candidate already, we've determined that we can reach up to X-Y percent for you. And while salary would normally be between $X-$Y, we can stretch to $X-$Y. This is above anything we had anticipated, but for the best of the best, we're willing to do what's necessary. Below, we've outlined the inputs that went into this in greater detail.

1: Market Comps
We've done research in conjunction with [compensation firm] as to the market comp for this role. Attached, you can find those outlines for [role] within SF/NYC Tech Companies with up to $100M in revenue and up to 1,000 employees.

1. Equity
    a. According to Option Impact:
        i. X percent (50th percentile) to Y percent in equity (75th percentile)
    b. According to Radford Data:
        i. Ongoing Actual Grants Face Value of $X (50th Percentile) to $Y (75th Percentile)
        ii. For reference, a X percent to Y percent offer from Bolt would be worth roughly $X to $Y
2. Salary
    a. According to Radford Data:
        i. $X (50th percentile) to $Y (75th percentile) in total salary (base plus incentives)

2: Present Day Value
Typically, when we're hiring an executive from a major corporation, our present day value is lower than what they will make over four years at their company. However, the executive takes a bet on the upside.

Today's present day value is already higher than what you'd make at [Current Company] (excluding raises, new roles.... which I agree could be far larger). At X percent, the present day value would roughly be $YM * Z% = $Valuc. That—plus a $X a year salary—would put the present day value at $Y vs. roughly $Z at [Current Co]. I expected our present day value to be lower, so this number being higher out of the gate is a strong foundation. This assumes no further appreciation or increases in valuation.

3: Upside Case
On top of present day value, here's the upside picture of what this equity would be worth as we begin to surpass comparable companies in the space. In addition to larger, well-known companies like Stripe, PayPal, and Shopify, we've included lesser-known players that are more likely to be surpassable in the short term. For additional reference, our latest priced investor round back in [Date] was completed at a $XM valuation, although we have signed a term sheet recently at $X, which is closing imminently. (Please keep this information of the new round extra confidential.)

Comparable Company	Market Value (as of Aug 26, 2020)	Price per share at respective market value	Equity package value at respective market value
PayPal	$237B	$1,592.74	$X
Shopify	$125B	$840.05	$X
Adyen	$42B	$282.26	$X
Stripe	$35B	$235.22	$X
AfterPay	$26B	$174.73	$X
Coinbase	$8B	$53.76	$X
Bigcommerce	$6.4B	$43.01	$X
Klarna	$5.5B	$36.96	$X
Checkout.com	$5.5B	$36.96	$X

Funding (Extra Context)

I also want to give you some context on the caliber of our backers. We've partnered with top-notch backers to raise $140 million so far at very favorable terms with another $100 million closing soon. Some highlighted investors include:

- **Notable industry leaders:**
  - Laurence Tosi (former CFO Airbnb, former CFO Blackstone), Michael Vaughan (former COO at Venmo), Gary Sheinbaum (CEO of Tommy Hilfiger North America), Jon McNeil (board member at Lululemon, former COO at Lyft, and former president at Tesla), Mark Lenhard (former SVP strategy at PayPal), and Jonathan Weiner (co-founder of Shoptalk and Money2020)

- **Notable founders:**
  - Tom Proulx (co-founder of Intuit), Michael Baum (founder of Splunk), Jeff Fluhr (founder of Stubhub), and Michael Arrington (founder of TechCrunch) co-founder of Shoptalk and Money2020)

- **Notable firms:**
  - General Atlantic (best growth investor there is), Westcap, Tribe Capital, Founders Fund, Floodgate, Green Oaks, Activant Capital, Stanford's StartX Venture Fund, Streamlined Ventures, Glynn Capital, and Human Capital

- **Notable retail backers:**
  - Executives at AllBirds, Guess, Revolve, Crocs, ABInBev, Forever21, Peloton, Jet.com, Bombas, Stance, and Nike

Note: These are executives who personally invested in Bolt with deep conviction in our vision for democratizing commerce and a commitment to championing our product at their respective companies.

You're exceptional and we had never contemplated an offer this high.

However, we believe that, if we move forward, you're more than worth it and have the ability to really take Bolt to the next level. I want you to step in feeling 1000 percent committed to Bolt for the long term—and that is the motivation behind presenting an offer in this range.

Note that this is not a formal offer; we still have a handful of candidates we're considering, and we still have some final steps to conduct before making a formal offer. But please take this as a strong indication of our optimism and my personal excitement about working together.

Talk soon,

Ryan

This letter was to a major executive who ultimately did not accept our offer. You'll notice that the candidate was fearful of startup risk, meaning they were likely not a good fit in the first place. We likely should not have pushed this hard.

Hi _______,

Thank you for the honest conversation last night.

You told me that your heart says Bolt, but your mind says don't jump. All the gurus who study the universe are unanimous around one fundamental truth: Follow your heart. It is your guide to your divine path. Jump.

Security for your family is paramount. Your parents immigrated to this country without a penny to their name and worked their way up from nothing. We cannot ignore our pasts, and I respect the journey you and your family have been on. In fact, I admire it greatly. You started with nothing and are now one of the most desired executives on the planet. Pretty spectacular.

Out of respect for this, we will provide you with an even stronger cash foundation of $__ per year in base salary. Even though it is not the same as 40-year-old [larger company] stock, the combination of security and upside here is extraordinary.

In regards to Bolt's liquidity, we have had a thriving secondaries market for years. It is always at a premium to the last round (most other companies are at a discount). For instance, I just got wind of $15M in Bolt stock being sold at a $___B valuation marker.

What about macro risks? For one, Bolt is Covid-proof. This pandemic pushes everything to digital. When Covid hit, our volumes nearly doubled overnight. Furthermore, even other economic downturns will lead to businesses needing to digitize, streamline, and optimize conversion. Like Amazon, Bolt thrives no matter the macroeconomic climate.

The large company roles will always be there. As I've spoken to more and more people about you, I've realized how spectacular your reputation is. I can say with certainty that you have earned your stripes and will always be able to put food on the table. Any company will take you at any time.

But Bolt might only come around once in a lifetime. You said you regret missing out on [other fast growing fintech startup]. The universe has delivered you something even greater: Bolt. We have the opportunity to be 10 times as large as [that company]. You holding out at that point led you to something even better. But it is unlikely to get better than Bolt.

With Bolt, you can change the course of your family's wealth for generations. Our worst flatlining outcome is on par with your current [large company] offer. But, we have $100B in GMV coming online, dozens of partnerships unlocking, and so much momentum I find it hard to believe that flatlining will be the outcome. From my perspective, the likely case is 10 to 100 times that outcome. This will forever change your life and the lives of your loved ones.

There is no founder that will work harder for you. Period. You have my full and complete commitment to making this company and partnership everything we know it can be. The same goes for the full executive team. We have hired the best in the world who are all value-aligned and also ALL IN.

This challenge is what you're made for. You are going to learn an immense amount, no doubt. But there's also no doubt that you are going to crush it. You are a one in a billion leader. I truly believe that. I cannot wait to see [name] unleashed.

Take your time with this decision. Enjoy the chats with [the other execs]. And talk soon, my friend.

Ryan

THE MASTERPIECE OFFER LETTER

As my Bolt talent team will tell you, I really care about the design of our offer letter.

The offer letter is the document that a candidate will look at most when evaluating your company. It's what they will share with their significant other.

I put a crazy amount of attention to detail in our offer letter design.

Here is how our offer letter leads in:

Date

Dear [First Name]

Congratulations! We are excited to invite you to join the Bolt team to build the future of commerce. On behalf of the entire Bolt team, welcome.

You are clearly exceptional. What really excites us is your [LIST THIS CANDIDATE'S SUPERPOWERS]. We aspire to find individuals like you who have both the desire and the ability to push the envelope to build the future.

We are redefining the digital commerce space. It will be hard and there are a ton of challenges facing us. However, with the right mindset, strategy, and the world-class team that we're building, we know we can push the world forward.

Most importantly, we're going to have a lot of fun along the way.

Let's do this thing.

Ryan & The Bolt Team

The next sections outline the actual offer mechanics alongside a few sentences on our compensation philosophy:

OUR COMPENSATION PHILOSOPHY

Our compensation offer is designed to be high-market or above market. In addition, cash and equity compensation is subject to grow based on performance reviews.

We also offer a variety of benefits as part of your total compensation package. We cover 100 percent (75 percent for dependents) of medical, vision, and dental benefits. We offer flexible paid time off as well as a variety of generous additional benefits. Importantly, we want to offer you a meaningful ownership stake in Bolt.

OUR EQUITY PHILOSOPHY

At Bolt, we believe that the best outcomes happen when the team has true ownership and can participate in the upside. For this reason we consider equity compensation perhaps the most important part of the total package, on par with personal growth and the company's mission of democratizing commerce.

Even if your compensation philosophy differs from the above, be clear about what it is to the candidate. Clarity and transparency go a long way.

Finally, we list hypothetical equity outcomes. We used to have a slider that showed how much someone's equity could be worth at a valuation of $X, $Y, or $Z.

However, when your company is tiny and you're writing major numbers in the letter, it makes those figures seem completely unachievable. When you write valuations of comparable companies, it makes the figures seem more realistic. An example:

Comparable Company	Market Value (as of June, 2021)	Bolt Equity Package Value at respective market value
PayPal	322,000,000,000	$XX,XXX,XXX
Shopify	163,000,000,000	$XX,XXX,XXX
Stripe	95,000,000,000	$XX,XXX,XXX
Adyen	71,000,000,000	$X,XXX,XXX
Coinbase	62,000,000,000	$X,XXX,XXX
Klarna	45,600,000,000	$X,XXX,XXX
Checkout.com	15,000,000,000	$X,XXX,XXX
Affirm	18,000,000,000	$X,XXX,XXX
Bigcommerce	4,500,000,000	$XXX,XXX

Here you must add the appropriate disclaimers—that these numbers do not include dilution, which may be significant, as well as the fact that no outcome is guaranteed and there is a risk that the company will lose all value and their equity will go to zero.

Finally, add a section to the letter explaining why you believe your business has the upside potential. We have a one-page memo that outlines this. Don't assume candidates got all of this information during the interview process. Remind them of your reasons for confidence here.

At the end of an offer letter, list notable backers and any other credentials where possible.
Your offer letter is effectively a recap of your entire sales process. Vision, Product, Culture, Credentials: all of that should be in here.

THE DEAL ISN'T SEALED UNTIL THEY UPDATE THEIR LINKEDIN PROFILE

Even when a candidate has signed an offer letter, sometimes they'll back out before their start date. A good rule of thumb is that a candidate should not be considered hired until they have been at your company for one week. Another is that they have not truly joined the team until they have updated their LinkedIn profile. This means the new employee is committed enough to your company to broadcast the decision to their entire professional network.

If a candidate drops out, it means they were likely not a good fit for your company. On the other hand, perhaps their experience post-acceptance fell flat.

It's on you to make sure that doesn't happen. Imagine getting an offer of acceptance and then waiting in silence until the day you start. You and your team should make the candidate feel the love not just during the interview process but even after they accept. The relationship has only just begun!

HAVE EVERYONE WHO INTERVIEWED A CANDIDATE CONGRATULATE THEM

There is no better way to make a new hire feel welcome at your company than by having everyone who was a part of their interview and hiring process congratulate them. Even if the team has already sent notes to a new hire after the offer has been made, there's no harm in a second note!

We also like to send an email or Slack post to the company celebrating each new hire. That message consists of a one- or two-paragraph explanation of why the candidate is ridiculously awesome—and we include a note asking the candidate's future team members to contact them, then let us know when that has actually happened.

FOLLOW UP WITH THE CANDIDATE'S REFERENCES

I like to write personalized notes to the candidate's references thanking them for their time and letting them know that the candidate is joining Bolt. This is an important way to express gratitude, show that you value follow-up, and build another touchpoint with that reference, who might end up being a network node or a candidate in the future.

This will also cause the candidate's references to congratulate them, thus furthering buy-in.

CELEBRATORY DRINKS OR DINNER

Have you or someone on your team get drinks or a meal with the candidate between the time they've accepted the offer and the time they've joined your company.

If their spouse would like to be included, I highly recommend including them.

HAVE INVESTORS SEND CONGRATULATORY NOTES

This is similar to other touchpoints, but can often have a greater impact. It's one thing for you and your team to talk about how great your company is—it's another when someone who has put money into you is doing it. That's because that person has (usually) placed other bets, so candidates often won't expect them to reach out. It shows a full, company-wide commitment to the candidate—all the way up to an investor.

INVITE THE CANDIDATE TO ANY UPCOMING TEAM EVENTS

If you've followed the guidance above and you've had multiple touchpoints with a candidate—and feel they're more likely than not to join—go ahead and invite them to upcoming team events. It'll paint a vivid picture of your company's esprit de corps. It's also a

good way of making the new hire feel like they're already in the room with you as a full-fledged teammate.
This also spreads the effort of closing onto other team members. If you've been pushing for this candidate, then they may have heard from you enough already. At a casual dinner or a hackathon or a celebration of a product launch, the rest of your team can talk about all the great features of your company—and further convince the candidate that they ought to join.

ALWAYS LEAVE A POSITIVE IMPRESSION

Even if an interview process is going terribly, that person still matters. Leave a great impression on everyone who crosses paths with your company. If you've done your job right, even if a candidate is rejected, they'll be singing your praises.

There was no better example of this than Bolt's Head of Digital Marketing hire. Let's call him Jim. Jim's husband, let's call him John, had recently interviewed at Bolt for the role but it had been decided that he was too junior for the scope of the role. Unknowingly, a recruiter reached out to Jim, John's husband, for the same role! Jim went to his husband to show him the outreach email as a joke. But, John told Jim "No seriously, you should go interview. The company is amazing and I actually was too junior for the role. It's a much better fit for you." Weeks later, we hired Jim.

You may not always have a perfect hit rate here—you're human, after all. It's not possible to always leave a positive impression. But that should always be the intent throughout the process.

7

CONCLUSION

Recruiting is one of the most important things you'll do as a founder or leader. That's why over-investing in it matters so much. I hope this book will help you throughout the recruiting process, but if you take away nothing else, remember that recruiting shouldn't fall entirely on your own shoulders. Throughout this book, I've tried to highlight how your team, your network, your investors, outside recruiters, and anyone in your orbit can be helpful in finding and closing the best talent in the world.

Hopefully some of the tactical material in these pages helps you find great people—but the principles are what matter most. Your people become your company, and because of that, there's nothing you can do that will push your company forward more powerfully than mastering recruiting.

So go forth and recruit. Build a great team—and a great culture.

8

BONUS: RECRUITERS REVEAL THEIR SECRETS

I mentioned that I find professional recruiters to be an important part of the recruiting arsenal. They aren't just people who find great talent—they spend all day *obsessing* about great talent. So I thought I'd get a few of the best to share their advice on finding the highest-quality talent in the world. Without further ado...

JENNIFER CHRISTIE, CHIEF PEOPLE OFFICER @ TWITTER

My entry into HR was recruiting. My first hiring manager was the President of the United States, when I was working in the Office of Presidential Personnel. I moved on from there to an executive search firm and then to the opportunity to reimagine the executive recruitment function at American Express.

I have since held other kinds of HR roles and am now serving as a Chief People Officer, but my passion remains in recruiting. Why? Because I learned early in my career that finding the right people is the most important thing. Whether it is in public service or the private sector, talent unlocks the potential, purpose, and vision of an organization.

Here are a few things I have learned along the way that I think create first-rate recruiting organizations and help land the best talent.

1. **Nail the Candidate Experience —** The candidate experience is everything! From the second a candidate starts contemplating your organization as the next stop on their career journey, they should feel informed, respected, excited, and valued. This means intentional care is directed at any marketing materials and job sites, so they are informative and reflect your culture and values. It means that at every step of the process they receive prompt replies and updates, have their questions answered, and get a full, honest picture of the role and the organization. It means you are being thoughtful about how they spend their time. Are you having candidates meet with everyone who wants but might not need to be part of the decision? Instead, are you protecting the candidate's time and only setting up meetings that are critical to get to a decision AND inform THEM about the opportunity? People have choices, and they will be assessing YOU with every interaction and communication.

   The experiences that unsuccessful candidates have is arguably more important than the experiences of the people you hire. The people you hire will naturally have a more positive view of the company, but for every one person you hire, you decline MANY more. That ONE person you hire will probably say great things, but what about all those others who didn't get an offer? What will they say? It matters because this group is exponentially bigger than the group you will hire, so they will be a bigger voice in building your talent brand. Were they treated with respect or were they ghosted? Did they get a timely decline or were they strung along? Did they receive usable feedback to help them improve their chances for the next role or were they given generic messages about why they didn't move forward? The talent brand you build will be a key factor in whether people you want to hire will even apply to work for your organization.

2. **Do the Prep Work —** Make sure you know what qualifications and experiences you are looking for before you hit the market. This seems basic, but too many hiring managers try the 'I will know what I need when I see it' approach. Establishing a clear job description and hiring rubric to clarify the 'must haves' and 'nice to haves' is critical. Not only does this help your recruiters be efficient and effective, but also it keeps you from wasting the time of people who are not right for the role in the first place. It also helps eliminate bias from the process at every step along the way.

3. **Look for a Growth Mindset —** If a candidate seeking a specific role 'checks all the boxes' of skills and experience, that is usually a red flag for me. I seek people who crave development and will stretch themselves to learn. People who have the courage to take risks and the resilience to fail and move forward are the ones I want to hire. Doing a job again that you have already nailed before is the safe, and I would even suggest, lazy route.

4. **Don't Under-level or Under-pay —** Under-leveling someone coming in the door is a bad practice. So is trying to hire someone for less than the market rate for their skill set and experience and the responsibilities they will have. Maybe you are worried that it will mess up your org design or cause issues with others on the team if you bring someone in at a certain level or pay. Maybe you think you can save some money. If those are issues, fix the org design or levels or pay of people on the team. Don't fix it by under-leveling or under-paying a new hire.

   You may be able to get them in the door, but they will figure out quickly how they are stacking up against their peers. Also, with pay transparency continuing to evolve, they will have more visibility to what they should be paid. If they feel they are under-leveled or under-paid, dissatisfaction will set in quickly and trust will be broken. Those first months are the most critical in retaining your new hire, and it is hard to turn around a relationship that starts out this way.

VIVEK PUNJABI, HEAD OF TALENT @ BOLT

Bolt's vision to build an iconic consumer brand has been driven by hiring and retaining the best talent in the world. We have embraced this challenge head on. Effective hiring is a byproduct of trusting the brand. This starts before candidates engage with Bolt. For example, I like sneakers. When it is time for a new pair, Nike is my first choice. I grew up seeing their logo, billboards, and commercials everywhere. I see my friends and family always wearing Nike shoes. Naturally I want them, too.

Recruiting is similar: the more trust and credibility we are able to build with a candidate before they engage in the recruiting process, the better our odds. That's why we've hired a top-tier marketing organization, open sourced our culture (conscious.org), and posted about every single hire we make on LinkedIn to celebrate our victories. We are constantly updating the world about what we are up to, a strategy that has attracted some of the best talent in the universe.

Branding is the initial step in creating a stellar candidate experience. This concept is layered with a World Class Talent Organization. We believe in 21st century recruiters possessing deep business and domain knowledge. Our team knows how to engage, sell, and ultimately close A+ candidates against multiple other opportunities. They are the face of the company acting as gatekeepers to maintain the level of excellence that we expect. They keep an active eye on adding to our culture and ensure we don't hire hastily.

Being all-in on a hire is a key to success. When hiring, we often ask ourselves, "Would we want to report to this person?", "Would we want to work with 100 more people like this?" We look for an astounding, unvarnished YES here. To determine this, we've looked for candidates with a combination of persistence and curiosity. This matching of grit and complexity has been the core DNA in which we've hired many of our top folks at Bolt. It takes discipline and patience to hire well, but if done correctly can be incredibly rewarding.

Ideally, you master how to do every part of the recruitment process as Ryan talks through in this book, and, if you do, you'll be able to get to my favorite part of recruiting....the close! Through all of the multiple touchpoints, you've finally landed on your dream candidate. Now it's time to "strike to close" — a term we've coined that means you've done such an excellent job throughout your recruitment process that a candidate feels like there is no other choice but to join your company. Even when they are presented with multiple other opportunities, they've fallen so in love with what you have to offer that closing is no longer a long drawn out process, but more of a formality that asks a candidate to simply sign on the dotted line. They are blinded by everyone else but you.

TIDO CARRIERO, CHIEF PRODUCT & ENGINEERING OFFICER @ SEGMENT

One of the most common mistakes I see people making while hiring is what I refer to as "brand-itis." It's an easy mistake to make: I should simply go find the biggest hire based on traditional metrics like company brand, revenue scale, team size, title, etc. While many of these are helpful signals, none of them actually predicts success nearly as well as answering two questions:

Y-intercept: Can the person do the job I need them to today today?
Slope: Does this person have a track record that indicates that they will be able to scale into 2-3 more jobs that you'll have for them as the company scales?

So how to actually figure that out? Almost every interview loop I build has two key questions — a "breadth interview" used for efficiently screening candidates at the beginning of the process and a "depth interview" used for digging deep during the onsite. Below is a synopsis of each.

BREADTH INTERVIEW

I like to really focus on the past two jobs that the candidate has had. What I'm really trying to understand is the deeper context behind the last two roles. I want to know what things were like at the beginning of their job:

- What was their mission given to them by their manager?
- What was the team size?
- What was the revenue base?
- What was the vibe on the team?
- Why did they take the job?

And of course this is just background context I need to answer the questions I'm really curious about:

- How did the company grow during that time (team, revenue, etc.)?

- How do they describe the overall impact they personally had at the company toward this? What specific projects drove the impact?
- How did their role change over time? (Promotion(s)? Bigger team given to them to manage? Harder projects with more impact?)

I find that with some calibration on a role, I can roughly get to 80 percent confidence about whether or not they will be a great fit for the role in a 30-minute screen, with some time for selling. (Maybe you'll need more like 45 to 60 minutes if you're a less experienced hiring manager or doing your first few interviews for the role and getting calibrated.)

DEPTH INTERVIEW

This interview is really about validating some of the stories you heard in the breadth interview. My default question here for candidates is to have them design a presentation. I ask candidates to spend no more than three hours (absolute max: five hours!) to present on a topic in more depth. I get very clear on what I'm looking for. Here's a few examples of what I might ask for different roles:

- **Product Leader:** I really want to understand how you collaborated with your GTM partners. Can you explain in detail how this collaboration worked, especially in an area that had more tension?

- **Engineering Leader:** I really want to hear about a time you made a foundational bet that really paid off down the line. Can you explain in detail the decision at the time and why the foundational bet was the right one? (I also want to hear an example of a time this didn't go correctly and your retrospective on that.)

- **Sales Leader:** I really want to hear about how you've driven sales efficiency outcomes. Walk me through the specific initiatives you drove, why you picked those, and the overall metrics you were able to drive

Note that a presentation isn't always the right format for the depth interview. For instance, if we're interviewing an IC software engineer, I'd almost certainly make the depth interview a pair programming exercise, which is a much better way for them to showcase their work. However, the principles still stand: I'll tell the candidate what I'm hoping to learn from this interview, how they will be graded, and give them every opportunity to put their best foot forward.

A FEW FINAL NOTES

- While I'd say the above two interviews really predict things most of the time, there are other interviews that are important — e.g., you'll definitely want a culture interview (with a real rubric that actually measures based on values. Countless studies show that "would I want to grab a beer with X?" is a great way to ensure a lack of DEI outcomes.)

- For the two concepts—y-intercept (can the person do the role we need today?) and slope (has this person crushed every role they've been in and done the next level role?)—you should calibrate to what your company needs. In my experience, I've found that slope tends to be much more important than y-intercept, so keep this in mind, especially as you are doing the breadth interview.

- References are a good way to validate your findings from the breadth interview. If this is an option, you should take it every time (backdoor references are best, if you can manage it!).

- Never forget: you are selling at every step of the way. Even candidates I've passed on have come back to thank me for being an incredible mentor and partner to them throughout the process — this is key to building a world-class hiring brand.

CONSCIOUS.ORG

This book explains how to recruit. Ready to build? Check out **Conscious.org**, a framework for bridging humanity with execution in the workplace.

FUNDRAISING

Ready to Fundraise? Check out Fundraising, Ryan's book on Fundraising.

SPREADING THE WORD

Ryan's books are all self-published and grassroots. If you found this helpful, please leave a review on Amazon and post that review to your Twitter or LinkedIn!

ABOUT RYAN BRESLOW

Ryan Breslow is the founder and CEO of Bolt, the technology company democratizing commerce by bringing lightning-fast, one-click checkout to the world.

Ryan believes that culture is the next frontier for innovation. Together with Bolt, he launched the Conscious Culture movement (see conscious.org) in 2021 to open-source all of Bolt's cultural guidelines with the rest of the internet, inspiring other companies to bridge humanity with execution in the workplace.

In high school, Ryan got his start in commerce bagging groceries and building e-commerce sites for businesses of all sizes—from small mom-and-pop shops to large corporations. It was during this time that Ryan experienced first-hand the shortcomings in checkout and online commerce, inspiring him to design a better alternative.

He went on to attend Stanford's Computer Science program and co-found the Stanford Bitcoin Group, dedicating research to the future of money and financial systems. In 2014, he left Stanford early to create Bolt and pursue his vision for democratizing commerce by perfecting checkout.

Outside of Bolt, Ryan is also the founder of Eco, a digital global cryptocurrency platform that can be used as a payment tool for daily-use transactions.

In his free time, Ryan enjoys dancing and practicing yoga and mindfulness. He founded The Movement, a charity that provides free dance classes for underserved communities in Miami that may not otherwise engage with dance and the arts because of financial constraints.

Printed in Great Britain
by Amazon